Mel Smells!

Katherine M. Parisky

Illustrations by Sandy Parisky

Illustrations were painted with Dr. Ph. Martin's non-toxic liquid "Hydrus" fine artist watercolor pigments and "Pilot V" liquid ink pen on 90lb acid free cold press "Fabriano" watercolor paper.

SUMMARY
A tiny fruit fly named Mel tells us about the special way he "looks" for his food and along the way, finds he has an important job.

Audience: Ages 4 and up.

Search keywords
1. Science for children 2. Insects 3. Fruit flies
4. Life cycle 5. Pollen 6. Smell 7. Taste

For Liam, Nevi, and Bri

ISBN: 978-1-61150-045-5

Published by: Fruit Fly Press, Boston, MA

My name is Mel! That's short for Melanogaster.

Scientists call me by my full name, Drosophila melanogaster.

I am a very small **insect**.

My body is so small that you would need a microscope to see all of my parts. The book illustrator decided to draw me here much larger than I really am. Now you can see even the tiniest hairs that cover my body.

actual size
rice
fruit fly
¼ ½ 1 inch

About 2 weeks ago, I started out as a microscopic egg laid by my mother inside a piece of rotting fruit. (My mother can lay 400 eggs. Can you imagine having 400 sisters and brothers?)

After about 14 hours, I crawled out of my egg.

We baby fruit flies are called **larvae** and we don't look anything like our parents. We look more like tiny worms.

Over four days I wiggled through the rotting fruit, drinking up fruit juice. I got big and fat!

Then I built a house for myself called a **puparium** and I stayed there for another four days. Inside I changed from a teeny, squirming larva into a fully formed adult fruit fly.

This transformation is called metamorphosis.

When I was ready, I pushed myself out. I'm not a worm anymore.

Now I have two wings, six legs, and thousands of tiny hairs that cover my entire body!

Egg

Larva

Pupa

Adult

I wake up every morning before the sun is high and begin a treasure hunt for sweet smells.

When I smell something good, I go investigate.

You may find me buzz buzz...
buzzing around your kitchen sometime.

Smelling is very important for deciding what I like.

I have to smell my food before I decide that it is safe to taste it.

Since I don't have a nose like yours, I use my **antennae** and the very tiny hairs on my head to smell my food.

Do you like to smell your food before you taste it?

The tiny hairs on my head are called **sensory sensilla**. The sensilla contain nerves..

My sensory sensilla let me smell, taste, and feel vibrations.

When I smell something good, I land right on top of it. Then I use my sensory sensilla to taste test the food.

Do you like my table manners?

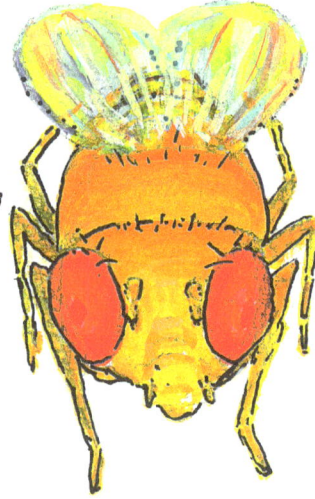

I like sweet foods the best! Fruits are my favorite. Of course that is why people call me the fruit fly.

My sensilla send important messages to my brain. The messages help me to decide if I want to stop and eat.

Are your favorite foods sweet or tart?

What I really like is rotting fruit! I love that yeasty-vinegar smell that comes from fermentation. You say, "Yuk!" I say, "Yum!"

What foods do you like to smell?

I love brown bruised bananas... I love mouthwatering mushy melons... I love squishy spoiled strawberries!

When my sensilla detect over-ripened fruit (or your compost pile), a quick message is sent to my brain, telling me to land, open my mouth, and drink up.

My mouth has a funny name. It's called a
proboscis. It works like a straw. I use my
proboscis to slurp up the fruit juices.

Sometimes I get fooled by the things that I smell.

There are flowers that pretend to be a tasty food.

One time the scent of a special lily flower called out to me.

It made itself smell like a piece of rotting fruit.

But when I went inside, there was no food!

Instead, all I found was yellow powder sticking to the
sensilla on my legs and wings.

It turns out that the flower had a job for me. My job was to carry that yellow powder (pollen) to other flowers.

When I carry the pollen, I help the flowers make more plants! More plants mean more fruit for me (and for you) to eat.

When I am tired from all my work, I like to find a cool, shady spot to rest. I know that when the sun is high, it is the right time for my **siesta**!

Do you have a favorite time of day to take a nap?

For more about the words and scientific terms related to Mel's story, see the glossary pages.

GLOSSARY

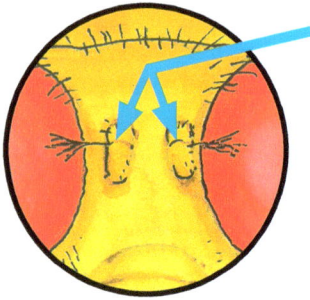

An-ten-nae
A pair of branch-like arms that stick out on the head of insects are called the antennae. Insects use their antennae to sense all sorts of things including: touch, movement, vibration (sound), temperature, taste, and smell.

Cre-pus-cu-lar
There is a special name for animals that are active at twilight. They are called crepuscular. This type of animal is most active just before sunrise (dawn) and again just after sunset (dusk). Crespuscular animals, like the fruit fly, sleep during the hottest time of the day.

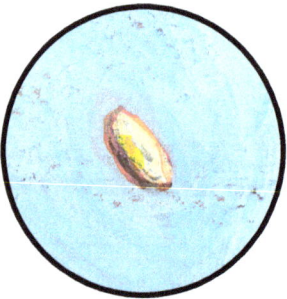

Em-bry-o
The embryo is the stage before an animal is ready to be born (or hatched). For fruit flies the embryonic stage lasts for about 22-24 hours.

In-sect

To be called an insect, an animal must have three segmented body parts: the head, thorax, and abdomen. All insects have six legs attached to their thorax. Most insects have wings, but some do not.

In-vert-e-brate

Invertebrates are a group of animals without bones! There are many different types of invertebrates. Invertebrates such as the jellyfish or worm have a fluid filled skeleton. Other invertebrates, like insects and crustaceans, have a hard outer shell called the cuticle.

Lar-va

Many insects go through a larval stage (when the animal looks very different from the adult). For example, at the beginning of a young fruit fly's life, just after it hatches from an egg, a larva looks like a worm. Plural lar-vae.

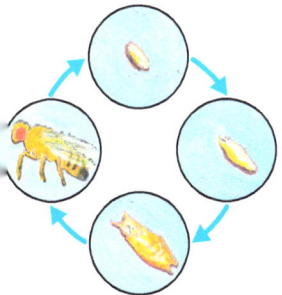

Life cycle

Every organism goes through a series of steps or changes in its development. From the beginning of life to death, a fruit fly goes from an egg to a larva, and then a pupa before becoming an adult.

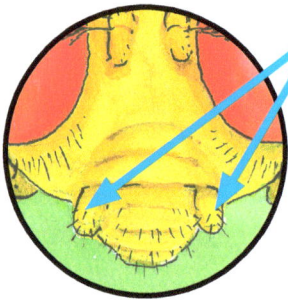

Max-ill-ary palp

The maxillary palp is a mouthpart that sticks out on the head of a fruit fly. The mouthpart is covered with specialized hairs called sensilla. The fruit fly uses its maxillary palp for smelling and for taste testing tiny samples to make sure it is safe before actually eating the food.

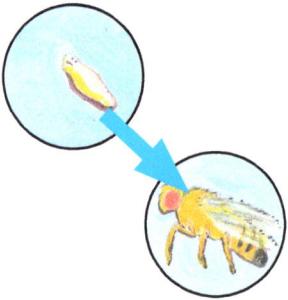

Meta-morph-o-sis

Metamorphosis is a big change that some animals go through before becoming an adult. Metamorphosis is a very common process in an insect's life cycle. At first an insect body may look very different (immature) before it changes into its adult form.

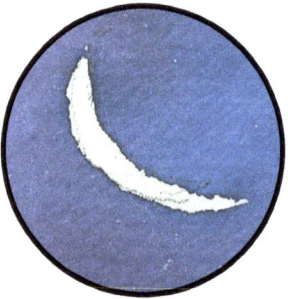

Noc-turn-al

At night, while you are sleeping, some animals are awake, busily exploring and eating! While it is too dark for our eyes to see very well without a flashlight, there are certain animals and insects that only search for food at night; they are called nocturnal. Even plants can tell the difference between day and night. During the day we see blooming flowers with pedals open wide but some flowers close their pedals at night to rest!

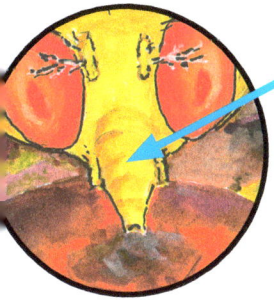

Pro-bos-cis
Some animals and insects have a long extension on their heads called a proboscis that they use for drinking.

Pu-par-ium
The house (hardened skin) that an insect uses to protect itself during the pupal stage of metamorphosis is called a puparium..

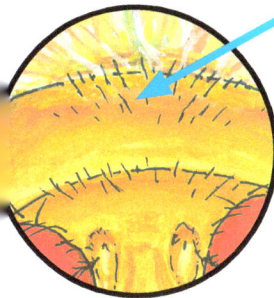

Sen-so-ry sen-sil-la
Small hairs on an insect are called sensory sensilla. Sensilla are sense organs; there are many types of sensilla found all over the insect body. There are different sensilla hairs for sensing touch, taste, smell, and temperature.

Si-es-ta
A nap at the hottest time of day (often after midday meal) is called a siesta.

www.ingramcontent.com/pod-product-compliance
Lightning Source LLC
Chambersburg PA
CBHW051559190326

41458CB00029B/6476